Multi-modal Partner 2.0.12 Tool:
Quick Start Guide
2012 Data Year - United States Version

United States
Environmental Protection
Agency

Multi-modal Partner 2.0.12 Tool: Quick Start Guide 2012 Data Year - United States Version

Transportation and Climate Division
Office of Transportation and Air Quality
U.S. Environmental Protection Agency

United States
Environmental Protection
Agency

Office of Transportation and Air Quality
EPA-420-B-13-027
June 2013

Table of Contents

What's New in this version?

Version 2.0.12 of the Multi-modal Tool contains a number of updates and revisions to the prior version, 2.0.11. These updates have been implemented to enhance the usefulness and clarity of the Tool, and are summarized below:

- Information regarding the split between U.S. and Canadian operations is now required.
- Checkmarks appear next to key steps on the **Home** screen once these steps are successfully completed.
- At least two separate contacts must be specified in the Contacts section.
- DOT Numbers and NAICS codes may now be entered for component fleets.
- Component Rail fleets may now be defined and characterized using the embedded Rail Tool, by selecting the Rail Mode under Step 3 on the **Home** screen. In previous versions rail activity was entered using the Logistics Mode.
- Partners can now provide information regarding their SmartWay program participation under Step 8 on the **Home** screen, "Provide Additional Information".
- Upon completion of the Tool the Partner generates and saves an "XML" file for submittal to EPA. The XML file is much smaller in size than the Excel file, which was previously sent to EPA.
- Screen demo icons are available on key screens. Selecting these icons will link the user to a video demonstrating the use of a given screen.
- Miscellaneous text and format updates for clarification.

Overview

In this guide you will learn about:

1) SmartWay Basic Information

2) Joining SmartWay Transport Partnership as a multi-modal carrier

3) Understanding the details of the SmartWay partnership agreement

4) Meeting software/hardware requirements for participating in the program

5) Gathering the data necessary for participation in SmartWay.

Please review this guide carefully BEFORE attempting to gather your company data, or entering data into the Multi-modal Tool. Understanding the basics of the program will simplify your SmartWay experience.

WARNING:

***Before beginning, use this chart to make sure you are choosing the right tool for your operations! ***

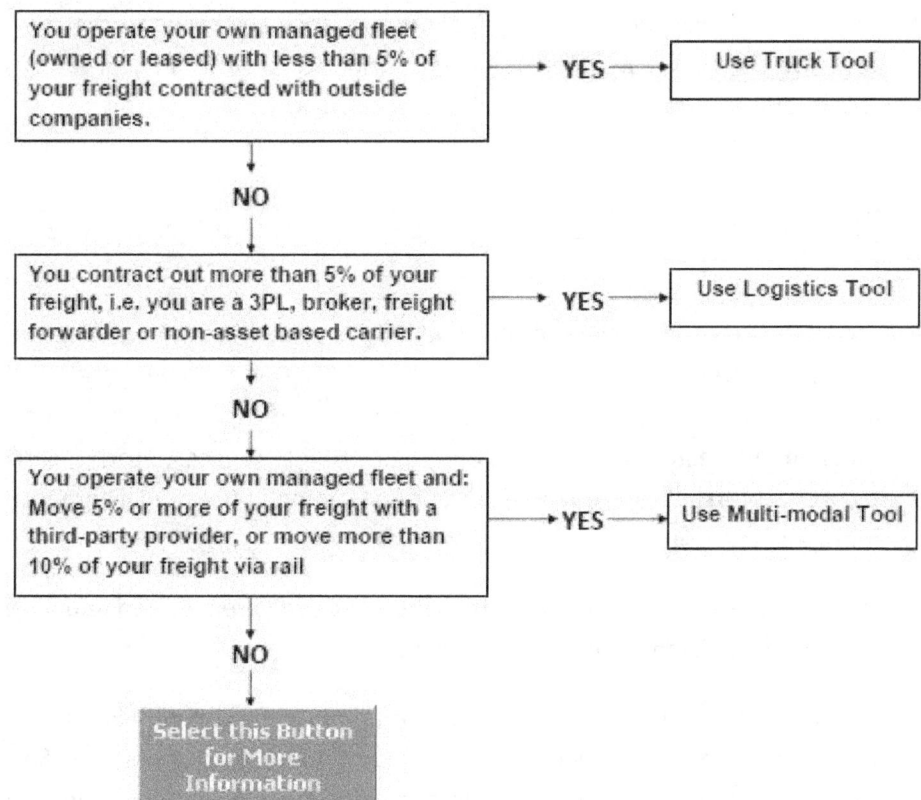

Is this the right SmartWay tool for me?

If none of the above statements is applicable, contact EPA SmartWay at 734-214-4647 for assistance.

Section 1—Basic Information for Multi-modal Carriers

This section covers frequently asked questions and essential information about the SmartWay Transport Partnership and how multi-modal carriers can participate.

WHAT IS THE SMARTWAY TRANSPORT PARTNERSHIP?

Launched in 2003, the SmartWay Transport Partnership is a public/private collaboration between the EPA and the freight industry to improve fuel efficiency, increase environmental performance, and encourage supply chain sustainability.

Five types of freight transport companies can join SmartWay.

- Freight shippers
- Logistics companies (including 3PLs/4PLs[1])
- Truck carriers
- Rail carriers
- Multi-modal carriers

Companies join the SmartWay Transport Partnership by submitting a partner tool to SmartWay. The SmartWay tools (1) assess freight operations, (2) calculate fuel consumption and carbon footprints, and (3) track fuel-efficiency and emission reductions. SmartWay tools must be submitted each year for the company to remain a partner in good standing.

SmartWay ranks Partners' efficiency and environmental performance and recognizes strong performance with access to the "SmartWay Partner" logo. Superior performance in various Partner categories is also recognized through the SmartWay Excellence Awards.

WHY DO MULTI-MODAL CARRIERS JOIN THE SMARTWAY TRANSPORT PARTNERSHIP?

The SmartWay Transport Partnership provides multi-modal carriers with ways to reduce the environmental impact of their freight operations and address costs. Designed <u>with and for</u> the freight sector, the SmartWay Transport Partnership delivers solutions to marketplace needs and challenges. With access to the latest in EPA-tested technologies and peer-provided success stories, carriers that join the SmartWay Transport Partnership can gain a better understanding of their environmental footprint and assert their corporate leadership.

 Additionally, SmartWay Partners are associated with an internationally recognized and respected brand that symbolizes cleaner, more efficient transportation choices.

Multi-modal carriers that join SmartWay move significant amounts of freight with SmartWay Carriers. The higher efficiency of carriers in SmartWay compared to non-SmartWay carriers gives SmartWay multi-modal carriers an opportunity to improve corporate freight efficiency, reduce their carbon footprint, and voluntarily advance freight sustainability for themselves and their customers.

---

[1] 3PLs/4PLs Third party logistics companies/fourth-party logistics companies.

Participation in SmartWay helps SmartWay Logistics Partners

- identify opportunities to improve efficiency,

- demonstrate efficiency to customers and stakeholders,

- evaluate and compare carrier performance, and

- determine the company's transportation carbon footprint.

SmartWay provides multi-modal carriers with free tools that help assess and compare various freight transportation options with detailed reports and analyses that support better business decision making.

HOW DO MULTI-MODAL CARRIERS JOIN THE SMARTWAY TRANSPORT PARTNERSHIP?

Multi-modal carriers join SmartWay by submitting a SmartWay Multi-modal Tool (hereafter known as the "Multi-modal Tool," or simply, the "Tool.")

Multi-modal carriers that submit tools that are approved by EPA are known as "SmartWay Multi-modal Carrier Partners."

When a multi-modal carrier submits a Multi-modal Tool to EPA, they agree to the requirements stipulated in the SmartWay Multi-modal Carrier Partnership Agreement--notably, that they will measure and report the emissions performance of their company _annually_ and provide supporting documentation to EPA upon request.

All SmartWay Multi-modal Carrier Partners agree to complete and submit the SmartWay Multi-modal Tool to:

- define fleet(s) composition
- characterize fleet activity
- individually benchmark multiple fleets
- track annual changes in performance

Upon approval of a Multi-modal Tool submission, a multi-modal carrier will be identified as a SmartWay Multi-Modal Carrier Partner on EPA's website, on the SmartWay Partner List, and in a database used to identify companies that meet SmartWay's annual requirements.

HOW DO I JOIN IF MY PARENT COMPANY HAS MULTIPLE FLEETS?

Companies that join the SmartWay partnership should include all of their fleets in their submission. If a company wishes to list multiple fleets in the Multi-modal Tool, they should list these fleets as their customers can hire them. Internal fleets invisible to a customer should not be listed separately. Companies will be listed at the Company level in the SmartWay Partner list on the SmartWay website, and each individually defined fleet will appear as a separate entity in the SmartWay Carrier Data file that customers use to identify which fleets they do business with in the Multi-modal Tool.

SmartWay highly recommends developing your list of fleets before beginning your data entry process. Any fleet that a shipper or logistics company could hire directly should be listed as a separate fleet in your Multi-modal Tool submission.

The best strategy is to have a clear idea of how to define your companies before filling out the tool.

WHAT DATA DO I NEED TO GATHER TO COMPLETE THE MULTI-MODAL TOOL?

To participate in SmartWay, multi-modal carriers need to gather the following essential information to complete the Multi-modal Tool:

- The official company name, EXACTLY as you would like it presented on the EPA website
- Company contact information
- Contact details for your primary contact
- Contact details for an executive contact (cannot be the same as the primary contact)
- Split between US/Canada operations
- Composite fleet details
    - Types of equipment,
    - Vehicle use by type
    - Fuel types used
    - Engine information, total ton-miles and total miles) plus:
    - Composite fleet names and associated component fleets
    - Component fleet allocation percentages
    - SCAC and/or Moto Carrier Numbers (MCNs) for composite fleets
    - Operation categories
    - Body Types
    - Business Unit Focus (e.g., logistics provider, freight forwarder, freight broker, or truck carrier) for logistics tool only
- Data sources for all data to be entered
- SmartWay ID number (if this is not your first tool submission)

This data must be provided for all of your company's fleets. This data reflects the amount of freight carried by each carrier, the distance that freight is carried, and the fuel consumed to carry the freight.

WHAT IS INCLUDED IN THE SMARTWAY MULTI-MODAL CARRIER PARTNERSHIP AGREEMENT?

To join the SmartWay Transport Partnership as a multi-modal carrier, you must agree to the language on the "Partnership Agreement for multi-modal carriers." When you begin working within the tool, you will be asked to check a box stating that you agree to the terms of the partnership agreement. **This agreement must be renewed annually.**

Please review this language with the appropriate personnel within your organization before completing or submitting a Tool to EPA. Your company's executive contact must approve this agreement.

Partnership Agreement for Multi-modal Carriers

With this agreement, your company joins EPA's SmartWay Transport Partnership and commits to:

1. Measure and report to EPA on an annual basis the environmental performance of your fleet(s) using EPA's SmartWay Multi-modal Tool. (Existing fleets must report the 12 months of data for the prior year ending December 31. Newly formed companies require a minimum of 3 months of operational data.)
2. Allow EPA to post performance results on the EPA SmartWay website/database.
3. Agree to submit supporting documentation to EPA for any data used to complete this Tool and agree to EPA audit of this data upon request by EPA.

In return, EPA commits to:

1. Promote company participation in the Partnership by posting Partner names on the EPA SmartWay website and in related educational, promotional, and media materials. EPA will obtain express written consent from the Partner before using the Partner's name other than in the context of increasing public awareness of its participation as described here.
2. Provide companies with industry-wide performance benchmark data as this data becomes available to EPA.[2]
3. Assist Partners in achieving emission and fuel usage reduction goals (subject to Federal Government Appropriations).

General Terms

1. If the Partner or EPA defaults upon this agreement at any point, the agreement shall be considered null and void.
2. Either party can terminate the agreement at any time without prior notification or penalties or any further obligation.
3. EPA agrees not to comment publicly regarding the withdrawal of specific partners.
4. EPA reserves the right to suspend or revoke Partner status for any Partner that fails to accomplish the specific actions to which it committed in the SmartWay Transport Partnership Agreement and subsequent annual Agreements.
5. The Partner agrees that it will not claim or imply that its participation in the SmartWay Transport Partnership constitutes EPA approval or endorsement of anything other than the Partner's commitment to the program. The Partner will not make statements or imply that EPA endorses the purchase or sale of the Partner's products and services or the views of the organization.
6. Submittal of this SmartWay Multi-modal Tool annually constitutes agreement to all terms in this Partnership Agreement. No separate agreement need be submitted.

---

[2] Individual corporate data will be treated as sensitive business information.

WHAT SOFTWARE AND HARDWARE IS REQUIRED FOR COMPLETING THE SMARTWAY MULTI-MODAL TOOL?

The Multi-modal Tool was designed in "Microsoft Excel Forms." Completing the Multi-modal Tool requires the following software and hardware:

- A 2003 or later version of Microsoft Excel

- Excel security level set at Medium or lower

- A PC running Windows XP or a newer operating system, or a Mac that is running the Windows XP operating system (the Tool does not currently work using the Mac operating system)

- A minimum of 10 megabytes of free disk space. More disk space may be required based on the number of companies you define in your tool

- Adequate memory (RAM) to run Microsoft Office

- A monitor resolution of at least 1,024 x 768[3]

Please check with the user guides for your computer, online support, or your company's IT department to make sure your system is set up to use the Multi-modal Tool.

We encourage you to make sure that you virus software is up to date, and to scan your PC before putting data in the Multi-modal Tool.

---

[3] The tool will also work at 800 x 600 resolution, but many of the screens will appear with scroll bars.

Section 2— Overview of Data Collection Requirements

This section will explore the data needed for completing the required sections of the Multi-modal Tool. **The Multi-modal Tool Data Entry and Troubleshooting Guide** explains more about the structure of the tool and the data entry process; this guide will focus primarily on the essentials for completing the tool.

INTRODUCTORY SCREENS

There are four screens that orient you to the tool: the partnership agreement, tool selection guidance, and data collection needs. These are general information screens; note that you MUST click the box indicating that you agree to the terms of the partnership agreement before moving on to the next screen.

PROVIDING US/CANADA OPERATIONS INFORMATION

To begin, you must specify the percentage of your fleet(s) that is licensed in the United States vs. the percentage that is licensed in Canada. If you operate exclusively in one country or the other, simply enter 100% for the appropriate country.

Next, enter the percent of your total fuel consumed that is attributable to United States-licensed vehicles and to Canadian-licensed vehicles. Then specify the data source used to determine your fuel consumption allocation. If your data source is not specified you may select "Other" and provide a text description of this source.

Finally, if you have operations in both countries, you must indicate if these vehicles use the same fuel mix in both countries. If the fuel mix is different, you must describe the difference in the types of fuel mix in the provided entry field. If you do not have the information to complete these fields, simply enter the approximate percentage of operations in each country and indicate that you use the same fuel type.

ESTABLISHING YOUR DATA COLLECTION YEAR

Before beginning your data collection, identify the last calendar year for which you have full annual (12 months) data. This means that you have data from January of the calendar year through December of the same year. If you are submitting for the first time and do not have a full year of operational data, please collect **_a minimum of three months' data_** for input into the SmartWay Tool. In your next update year, you will be required to submit a full year's data.

SECTION 1: SPECIFY OFFICIAL PARTNER NAME

Your Partner Name is the official name that your customers would recognize for your company—in other words, the name someone hiring you would look for.

You must specify you company's official Partner Name, exactly as you want it to appear on the SmartWay website.

For example, if you enter:

- ABC Company
- ABC Company, Inc.
- ABC COMPANY LLC

Your company will be listed **_exactly_** as you've entered above. Therefore, it is important to pay special attention to proper capitalization, abbreviations, annotations, and punctuation.

SECTION 2: ENTER CONTACT INFORMATION

The SmartWay tool asks for:

- **General company information** such as location, web address, phone number, etc.

- **A primary contact**[4] for any questions about your company's participation and tool submissions

- **An executive contact**[5] for participation in awards and recognition events

- **Additional contacts (optional):** Additional contacts may include anyone who is not the primary contact but may be involved with SmartWay (e.g., press/media contact, fleet manager, etc.).

Note that you MUST have at least two contacts listed in the contact information section of the Tool. SmartWay recommends developing an internal succession plan to make sure that your Multi-modal Tool submission schedule is maintained, in the event that a primary contact is reassigned, or leaves the company for any reason.

---

[4] The primary contact is the individual designated by the Executive Contact to directly interface with SmartWay regarding specific tasks involved in the timely submission of the tool. The Primary Contact is responsible for coordinating the assembly of information to complete/update company data; completing and updating the tool itself; maintaining direct communication with SmartWay; and keeping interested parties within the company apprised of relevant developments with SmartWay.) NOTE: To ensure that emails from SmartWay/EPA are not blocked, new primary contacts may need to add SmartWay/EPA to their preferred list of trusted sources.)

[5] The executive contact is the company executive who is responsible for agreeing to the requirements in the SmartWay Partnership Agreement, overseeing the Primary Contact (as appropriate), and ensuring the timely submission of the tool to SmartWay. The executive contact also represents the company at awards/recognition events. This person should be a Vice President or higher-level representative for the company.

Understanding Component and Composite Fleets:

Multi-modal companies are somewhat more complicated to characterize than truck or rail-only companies. Often, multi-modal services are comprised of a wide range of truck, rail, and/or dray services that may be provided by the company or purchased from third party providers by the company. To properly characterize the emission rates for multi-modal companies, one must account for this wide variety of operations and service levels.

> *Note: Multi-modal rail freight is defined as any freight transported by a multi-modal carrier, which may include intermodal containers as well as other cargo configurations such as boxcars, tanker cars, etc.*

To accomplish this SmartWay has developed the concept of "component" and "composite" fleets.

Composite fleets are entities that your customers can hire to move their freight. For example you may have a Truckload Division and an Intermodal Division, whereas **Component fleets** are organized around your internal management of the company.

In the Multi-modal Carrier Tool you will be asked to input data for all of your component fleets which will include those you own and manage as well as any hired services you purchase. Once these component fleets have been characterized you will be asked to build your composite fleets (the ones the public sees and can hire) from your component fleets. This process will attribute the appropriate weighted emission factors to your composite fleets, and demonstrate to your customers the benefits of using your multi-modal freight services.

The partner will then assign the appropriate operations of these component fleets to define the composite fleets that the public can hire:

Composite Fleet One: Truckload Division

  60% of its truckload fleet, 70% of its 3PL fleet

Composite Fleet Two: Intermodal Division

  40% of its truckload fleet, 30% of its 3PL fleet, 100% of its hired rail, 80% of its hired dray fleet, and 40% of its own dray fleet

Composite Fleet Three: Dray Division

  20% of its hired dray fleet, 60% of its own dray fleet

> *NOTE: Drayage operations should be included in the SmartWay Multi-modal Carrier Tool if your company controls drayage movements (e.g., receives payments directly to move the freight using your own drayage trucks, or purchases drayage services directly from a third party).*

Identifying Component Fleet:

On the **Home** screen select the pull-down menu next to "Choose a Carrier Mode" to display the available component fleet modes. The Multi-modal Carrier Tool currently allows users to define truck, logistics and rail component fleets. Other modes will be added to the Tool in the future, including air, and marine modules.

Note that some terminology on the Define Component Fleets screens are slightly inconsistent between the Truck Mode and the Logistics Mode, however the two modes work and act the same. For example, "fleets" within the Truck Mode are referred to as "Business Units" within the Logistics Mode.

When you select a Component Carrier Mode, a set of screens will automatically appear allowing you to characterize your fleet(s) for that mode.

There are four screens in the **Truck Fleet Characterization** section of the Tool. These screens are also common to the Logistics Mode:

1. Identify Fleets
2. Fleet Details
3. Operation Categories
4. Body Types

Screen 1: Identify Fleets is shown below for the Truck Mode below. Note that the data entry screens for Logistics Carriers contains the same required fields as for the Truck Mode, with the exception of the "95+% Control" field, as discussed below.

Steps for Completing "Identify Fleets" Screen

Using data collected using **Worksheet #2: Component Fleet Characterization** in the **Part II Workbook**, enter data for your first fleet:

- **Fleet Prefix (Partner Name):** Each of your component fleet names will begin with the name of your company. This fleet "prefix" will be whatever you enter in the Fleet Prefix (Partner Name) field on the Identify Fleets screen. By default, this field is automatically populated with the first 50 characters of the Partner Name that was entered on the Home screen. Whatever you enter for Fleet Prefix for the first fleet will automatically be used for any additional fleets you add. Similarly, any edits you make to the Fleet Prefix for the first fleet will automatically be reflected on each subsequent fleet. Note that this field is called Company Name for Logistics component fleets.

- **Fleet Identifier:** Please make sure to specify each fleet suffix name exactly as you want it displayed on the SmartWay website, including proper capitalization, any abbreviations, and punctuation. Remember that it will automatically be combined with the Fleet Prefix (Partner Name) field. NOTE: If you have only one fleet, you may leave the Fleet Suffix field blank, in which case your fleet name will simply be your Partner Name. Note that this field is called the Business Unit Identifier for Logistics component fleets.

Adding and Deleting Component Fleets

To enter another component fleet, select the **Add Another Fleet** button. To delete a component fleet, select the box next to the row you wish to delete, and then select the **Delete Checked Rows** button. Once you have confirmed or modified the Partner Name and specified the Fleet Identifiers, the full Fleet Names will be displayed on the screen to the right of the screen.

Adding Comments/Notes

Creating useful comments assists SmartWay Tool reviewers in approving your Tool as quickly as possible. Your comments help reviewers understand your business model. Any details that you can provide related to your business operations will speed up approval time.

 Please note the **ADD COMMENTS** button located at the bottom of the screen. This allows you to enter notes about the collection process, your assumptions and methods, data, or other information. These details could prove useful for your reviewer when you or someone else fills out the Tool next year. If comments have been added for a particular screen, the **ADD COMMENTS** button will be highlighted in yellow on your screen and will now read **VIEW/EDIT COMMENTS** to indicate to your reviewer that there are comments to be read.

A [HELP] button is also available should you need assistance. You will also notice small gray icons with question marks [?] displayed throughout the Tool. When clicked, these icons provide additional information about specific items located on the screen.

To proceed, select the <u>Fleet Details</u> tab at the top of the screen, or simply select the [NEXT] button at the bottom of the screen.

Steps for Completing "Fleet Details" Screen

Using data collected using **Worksheet #2: Component Fleet Characterization** in the **Part II Workbook**, enter data for your fleets. For each you will need to specify:

- **95% Control:** Select this box if your company controls over 95% of the operation of each fleet, weighted by miles. "Control" means that you operate/route the fleet, regardless of ownership status. Control includes dedicated fleets that you operate for other parties. If you contract out more than 5% of the fleet's operation, the Logistics mode should be selected instead for that fleet. (Note this box is not included in the Details screen for the Logistics Mode selection.)

  The key question regarding definition of "control" is: Am I able to influence the fuel efficiency of the trucks, and can I collect data on the trucks? SmartWay understands that control within the trucking business exists along a continuum. On one hand, some fleets purchase their own trucks, spec, maintain, and route the trucks, and have full operational control over the trucks, for example controlling the speed the trucks are allowed to drive, when and where they can idle, etc. These fleets have a high degree of control. On the other end of the continuum, some companies hire other parties to move the freight, and other than assigning a load with a pickup/destination point, have no interaction with the freight delivery, or ability to influence the fuel efficiency of the truck or the collection of data on the truck.

  If you can actively affect the fuel efficiency of the truck and collect the data necessary on that truck to include in this Tool, you have control. SmartWay understands that there are many organizational/operational permutations in the trucking industry and there may be "gray" areas that need further clarification. If you still have questions, you may contact your assigned SmartWay Partner Account Manager or the SmartWay help line at 734-214-4767.

- **North American Industry System Classification Codes (NAICS):** NAICS codes are the Federal standard for classifying businesses by activity type. NAICS codes can be from two or three up to six digits in length, and are organized hierarchically, with successive digits providing greater specificity in its description. Use the NAICS code(s) that most appropriately describes your company. Detailed information on NAICS codes can be found at http://www.census.gov/eos/www/naics/.

- **Standard Carrier Alpha Codes (SCACs):** The Standard Carrier Alpha Code is a unique 2-4 alphabetic character code used by the transportation industry to identify transportation companies. If your fleet has a SCAC, please input that information. If you have a single fleet that has multiple SCACs, enter all of the codes into the SCAC field, and separate them with commas. It is not required to enter SCAC information for each component fleet. SCACs are assigned by the National Motor Freight Traffic Association, Inc., (NMFTA). If you cannot remember your SCAC(s), please contact NMFTA before proceeding. You can find NMFTA contact information at http://www.nmfta.org/Pages/ContactUs.aspx.

- **Motor Carrier Number (MCN):** The Motor Carrier Number is a 6 or 7 digit number provided by the Federal Motor Carrier Safety Administration. If your fleet has a Motor Carrier Number, please input that information. It is not required to enter MCN information for each component fleet.

- **Department of Transportation (DOT) Numbers** are carrier identification number issued to all carriers in the U.S. by the Federal Motor Carrier Safety Administration, and can be up to 7 digits in length.

  NOTE: Entering SCACs, MCNs, and DOT numbers are optional; however, if you have this information you are encouraged to supply it to make sure that SmartWay Shippers and Logistics companies can identify you.

- **Fleet Type**: Fleet Type is defined as the service type for your fleet. There are two options accepted by the Tool—"For-Hire" and "Private/Dedicated." If your company has only one fleet, your "Fleet Type" selection will reflect your company's operations as a whole. If there are multiple fleets, each will have its fleet type defined separately.

- **Fleet Contact:** This contact should be one of the contacts you already identified in the Contact Information section as the contact for each fleet. NOTE: A drop-down menu in the Tool will supply this information; if there is a contact for the fleet that is not already listed in the Contacts worksheet, you will need to go back to the Company and Contacts screen to add the required contact information.

Steps for Completing "Operation Categories" Screen:

For each component fleet, fill out the **Operation Category (%)** information by indicating the percentage of operation on a mileage basis. Operational categories include:

- **Truckload (TL)** - Truckload shipping is the movement of large amounts of homogeneous cargo, generally the amount necessary to fill an entire semi-trailer or intermodal container. A truckload carrier is a trucking company that generally contracts an entire trailer-load to a single customer.

- **Less-than-truckload (LTL)** - Less-than-truckload carriers collect freight from various shippers and consolidate that freight onto enclosed trailers for linehaul to the delivering terminal or to a hub terminal where the freight will be further sorted and consolidated for additional linehauls.

- **Drayage** - Predominantly associated with port, or rail-head connections where freight is picked up, and moved to another transfer facility or transport mode terminal. Often these moves are short in nature, but can be longer depending on specific situations.

- **Package delivery (PD)** - Covers operations characterized by residential or business package delivery/pickup consisting primarily of single or small groups of packages. It does not include larger scale pickup delivery operations that are more properly characterized as LTL operations. Common examples of this type of operation are the brown UPS and white FedEx delivery vehicles.

- **Expedited** - Time-sensitive freight shipments, with trucks typically on stand-by.

Enter the percent of each operational category based on approximate mileage. This percentage calculation does not need to be exact but should be reasonably reflective of your fleet.

Steps for Completing "Body Types" Screen:

Next, fill out the **Body Type** fields, indicating the percentage by body type for each component fleet. Body Type categories include:

- Dry van
- Refrigerated (Reefer)
- Flatbed
- Tanker
- Intermodal chassis containers (pooled and owned)
- Heavy/Bulk hauler
- Auto carriers
- Moving
- Utility[6]
- Special hauler (e.g., Hopper, Livestock, and other specialized carriers)

---

[6] The utility category encompasses class 2b to 8b vehicles that do not carry typical commercial freight. Examples include garbage, recycle, service, work, dump, landscape, cement, bucket, boom, ambulance, armored, fire, farm, wrecker and other similar trucks. Because these trucks do not carry traditional freight payload, the user should self-define their payloads so as to make the emissions per payload efficiency useful to the user. SmartWay will not use the emissions per payload results for the utility category. Users may experience yellow or red warning labels on the Activity screen due to the unique nature of utility "payload." In the case of red alerts, simply input text defining your special conditions in the required text boxes that appear.

The percentages specified can be approximate, based on vehicle populations. The percentages for each fleet must sum to 100%.

Once you are sure your information is input correctly, you may click the **CREATE FLEET(S)** button at the bottom of the page.

If, at a point later in the data entry process, you realize that you need to add a new fleet or delete an existing fleet, you can return to the Identify Fleets screen. To add a new fleet, follow all of the instructions on the screen regarding defining your fleets, including clicking the **CREATE FLEET(S)** button. When you select this button, the system will create blank data entry forms only for the new fleet(s) you have added; the existing fleets will not be affected.

If you need to delete an existing fleet, simply check the box next to the fleet and then click the **Delete Checked Rows** button. Note that, if you have already generated data entry forms for the fleet you are deleting, the system will prompt you to confirm the deletion.

> *If you choose to delete a component fleet, and if you have allocated any activity to this fleet to the composite multi-modal fleets (defined under Step 6 on the Home screen) then you must re-allocate your composite fleet activity to reflect this change. Similarly, adding a new component fleet may require modifying your composite fleet activity allocations under Step 6 as well.*

As on the other screens there is a `HELP` button as well as an `ADD COMMENTS` button. Clicking `HOME` will take you back to the Home screen (see **Figure 24**). At this point you may define another component fleet following the same process, or proceed to Step 4 on the Home screen.

Data Requirements for Business Unit Focus Screen (Logistics Tool Only)

For each Logistics business unit you will need to specify the percent of total activity associated with the different Business Unit Focus categories, listed below.

- **Logistics Provider:** Logistics providers are non-asset based third parties that provide multiple, bundled logistics services. They may be involved in material management, transportation management, inbound and outbound freight, inventory management, 4PL activities, warehousing, cross dock, kitting, packaging, and sub assembly processes.

- **Freight Forwarder:** Freight forwarders are documentation specialists and consolidators of freight such as LTL (less than truckload) and LCL (less than container load). Freight forwarders normally provide pickup and delivery for domestic and international shipments, and provide the property transportation for a compensation or fee basis.

- **Freight Broker:** A freight broker buys and sells transportation services and normally works on behalf of a carrier or shipper.

- **Truck Carrier:** Truck carriers operate their own managed fleet (owned or leased). These fleets can be for-hire or private/dedicated. Refer to "Choosing the Right Tool for your Business Units" in Part I of this guide to determine if you should complete the SmartWay Truck Tool for the Truck Carrier portion of your operations.

Note: for each Rail fleet, you will also need to specify the Class of the associated fleet (1, 2, or 3).

SECTION 4: DOWNLOAD LATEST SMARTWAY CARRIER DATA FILE

To ensure that the Multi-modal Tool gives you the most current list of SmartWay carriers to select from, you must click the "Download Latest SmartWay Data Carrier Data File" button on the home screen. A new box will appear next to the button indicating the "Date of Current Carrier File." You are now ready to go on to selecting carriers and entering activity data for each fleet.

You will not be able to input the required logistics fleet data in Step 5 without this file. You must have an active Internet connection to perform this step.

S<small>ECTION</small> 5: D<small>ATA</small> E<small>NTRY FOR</small> F<small>LEETS</small>

Select Component Fleet for Data Entry

On the Home screen, you will now see all the fleets you created listed in the window below item # 5: Select Component Fleet for Data Entry.

There will be a status message after each fleet, indicating whether or not the data entry for that fleet is complete. The following information may appear beside a fleet name:

- **Not checked** - Data has not been entered yet.
- **Incomplete** - Some data is still missing and/or inconsistent.
- **Complete** - All data requirements have been met and validation has occurred.

In addition to the status indicators above, you may also see one of two qualifiers: "Errors" or "Warnings."

- **Errors** will prevent you from generating the **Internal Metrics Reports** under item #7, View Reports, and must be addressed before you can submit your Tool to EPA.
- **Warnings** will still allow you to run the **Internal Metrics Reports** and submit your data to EPA. However, it is strongly recommended that you carefully review each warning message before sending your data to EPA so that you can anticipate questions that may come from a Partner account manager (PAM) as a result of your data being outside the expected ranges. The method addressing errors and warnings is described for subsequent input screens in the following sections.

To add data to a particular fleet file, highlight the fleet name and then double-click. You will then proceed to the Tool **Data Entry Screens**.

S<small>ECTION</small> 5 D<small>ATA</small> E<small>NTRY</small>: E<small>NTER FUEL AND/OR ACTIVITY INFORMATION FOR EACH OF YOUR COMPONENT FLEETS</small>

P<small>LEASE REFER TO THE</small> T<small>RUCK</small>, L<small>OGISTICS, AND</small> R<small>AIL</small> T<small>OOL USER GUIDES FOR DETAILED INSTRUCTIONS REGARDING DATA ENTRY FOR YOUR COMPONENT FLEETS.</small>

SECTION 6 DATA ENTRY: DEFINE YOUR COMPOSITE FLEETS

Now that you have identified and characterized your component fleets, you will be asked to provide information for EACH composite fleet. In this section, you will be asked to allocate component fleet miles and ton-miles across the different composite fleets.

The "**Define Your Composite Fleet**" section of the Tool has three subsections:

1. Identify Composite Fleet
2. Allocate Component Fleet
3. Composite Fleet Details

The requirements for each subsection are described below.

Once you have entered data for all of your component fleets, and resolved any error notifications, you may proceed to Step 6, <u>Define Your Composite Fleets</u>. This step allows you to group your component fleets into larger, *composite fleets* operating across one or more modes. Composite fleets are entities that your customers can hire to move their freight. For example you may have a Truckload Division and an Intermodal Division for hire.

Identify Composite Fleets

First, enter the name of your composite fleets on the first screen, Identify Composite Fleets.Names include your Partner Name combined with your Fleet Identifier. Remember to enter your composite fleet names exactly as you want them to be seen on the SmartWay website.

> *Note: Your company's name and your composite fleet(s) will be listed on the SmartWay website to indicate your participation in the SmartWay Transport Partnership. Your shipper and logistics customers can also use the SmartWay Online Database to search for your company by the name you submit in the Tool, your SCAC number or your Motor Carrier Number. Therefore, it is <u>critical</u> that you identify your company and composite fleet(s) in the Tool as you would have them appear on the SmartWay website and within other SmartWay Tools.*

You may add more fleets by selecting **Add Another Composite Fleet**. To proceed to the next screen, select the Allocate Component Fleets tab or select NEXT.

Allocate Composite Fleets

Proceed to the Allocate Component Fleets screen to see a list all of the component fleets you defined under Step 3 (Define Your Component Fleets). All of the miles and ton-miles associated with each of these fleets must be assigned across one or more of the composite fleets you defined on the previous screen. You must allocate component fleet activity based on the percentage of total miles and total ton-miles attributable to each composite fleet. The total miles and ton-miles listed for each component fleet is automatically calculated from Step 5 where you entered the activity data for your component fleets. (Note that Total Miles refers to truckload miles for the Truck Mode, and railcar-miles for the Rail mode.)

Note that the percentage allocations must sum to 100 for each component fleet. In addition, if you enter a non-zero percentage for a particular total miles assignment, and zero percent for the corresponding ton-mile assignment (or vice versa), the Tool will display an error and you will not be allowed to proceed until reconciling the discrepancy.

The **Show Allocation** button located below the component fleet list allows you to view a summary of your composite fleet percentage allocations across all of your different component fleets, in order to confirm the accuracy of your assignments.

The **Validate Screen** button will validate the information you enter on this screen. The ratio of total allocated ton-miles to total miles is checked for each composite fleet. If any of these ratios differ from industry average payload standards, you will receive a warning message. These warning messages are intended to flag possible data entry errors; however these warning messages will not prevent you from submitting your completed Multi-modal Carrier Tool to EPA.[7]

Please refer to **Appendix A** for a description of the procedure used to calculate the gram per mile and gram per ton-mile performance metrics for composite fleets.

Select the **Return to Entry Form** button to leave the spreadsheet and return to the allocation screen.

---

[7] Validation warnings are issued for low payloads if the average payload for a composite fleet is < 14.5 tons (based on the out of range warning for Class 8b TL/Dry Vans – see Truck Tool Technical Documentation for details). Warnings are issued at the high end if the average composite fleet payloads are > 60 tons. This value was chosen based on the distribution of payloads reported by Multi-modal Partners in 2012, with approximately two thirds of Partners having payloads less than this cutoff.

Composite Fleet Details

First proceed by entering the SCAC and MCN values for each composite fleet if available (see **Figure 33**). Separate multiple values by a comma if necessary. These values are not mandatory but will assist Shippers and Logistics Companies in identifying your fleets. Next select the appropriate contact from the drop-down menu.

Once you have allocated all of your component fleets across the composite fleets, select the HOME button to return to the **Home** screen.

Appendix A: Worksheets for Data Collection

List of Worksheets

Worksheet #1: Company and Contact Information

#1. ENTER YOUR CONTACT INFORMATION:

General Company Contact Information

Company Name							
Headquarters Mailing Address							
City		State/Province		Zip		Country	
Main Phone Number		Toll-free Number		Cell number		Web Address	

Primary Contact Information

Primary Contact Name							
Primary Contact Mailing Address							
City		State/Province		Zip		Country	
Primary Contact Phone Number				Email Address			

Executive Contact Information

Executive Contact Name							
Executive Contact Mailing Address							
City		State/Province		Zip		Country	
Executive Contact Phone Number				Email Address			

Other Contact Information

Executive Contact Name							
Executive Contact Mailing Address							
City		State/Province		Zip		Country	
Executive Contact Phone Number				Email Address			
Contact's role in program							

Worksheet #2: Component Fleet Characterization

Complete this worksheet for <u>each component truck and/or logistic fleet</u> you will be submitting in the Multi-modal Carrier Tool.[8]

#2: Define your Fleets
Partner Name and Fleet Identifier

NAICS: _____ _____ **SCAC:** _____ **MCN:**_____ _____ _____ **DOT#:**_____ _____

FLEET TYPE:_____ _____ **95% Control*** _____

Fleet Contact:_____

Operation Category Percentages:
Truckload _____ LTL _____ Drayage _____ Package Delivery_____ Expedited_____

Body Type Percentages:
Dry Van_____ Reefer _____ Flatbed _____ Tanker _____ Chassis _____ Heavy Bulk _____
Auto Carrier _____ Moving _____ Utility _____ Special Hauler _____

* Applicable for Truck fleets, not Logistics fleets

---

[8] Component Rail fleets only require Partner Name, Fleet Identifier, NAICS code, Rail Class (1,2,or 3), and Fleet Contact.

Worksheet #3A: Identify Composite Fleets and Composite Fleet Details

Complete the following for each *composite* fleet you will be submitting in the Multi-modal Carrier Tool.

#3A: Define Your Composite Fleets
Company Name / Component Fleet Identifier

_____

SCAC: _____

Motor Carrier Number: _____

Fleet Contact:_____

Worksheet #3B: Allocate Component Fleets

Complete the following table for <u>each *component* fleet</u> you will be assigning to the composite fleets listed in Worksheet 3A.

#3B: Allocate Your Component Fleets

List each composite fleet identified in Worksheet 3A. Add lines to the table below or copy the table multiple times if necessary.

"% Miles" and "% Ton-Miles" columns must each sum to exactly 100% for each component fleet.

	Composite Fleet Name	% Miles	% Ton-Miles
1			
2			
3			
4			
5			
6			
7			
8			
9			
10			
11			
12			
13			
14			
15			
16			
17			
18			
19			
20			

www.ingramcontent.com/pod-product-compliance
Lightning Source LLC
Chambersburg PA
CBHW081416170526
45166CB00010B/3359

9781500606176